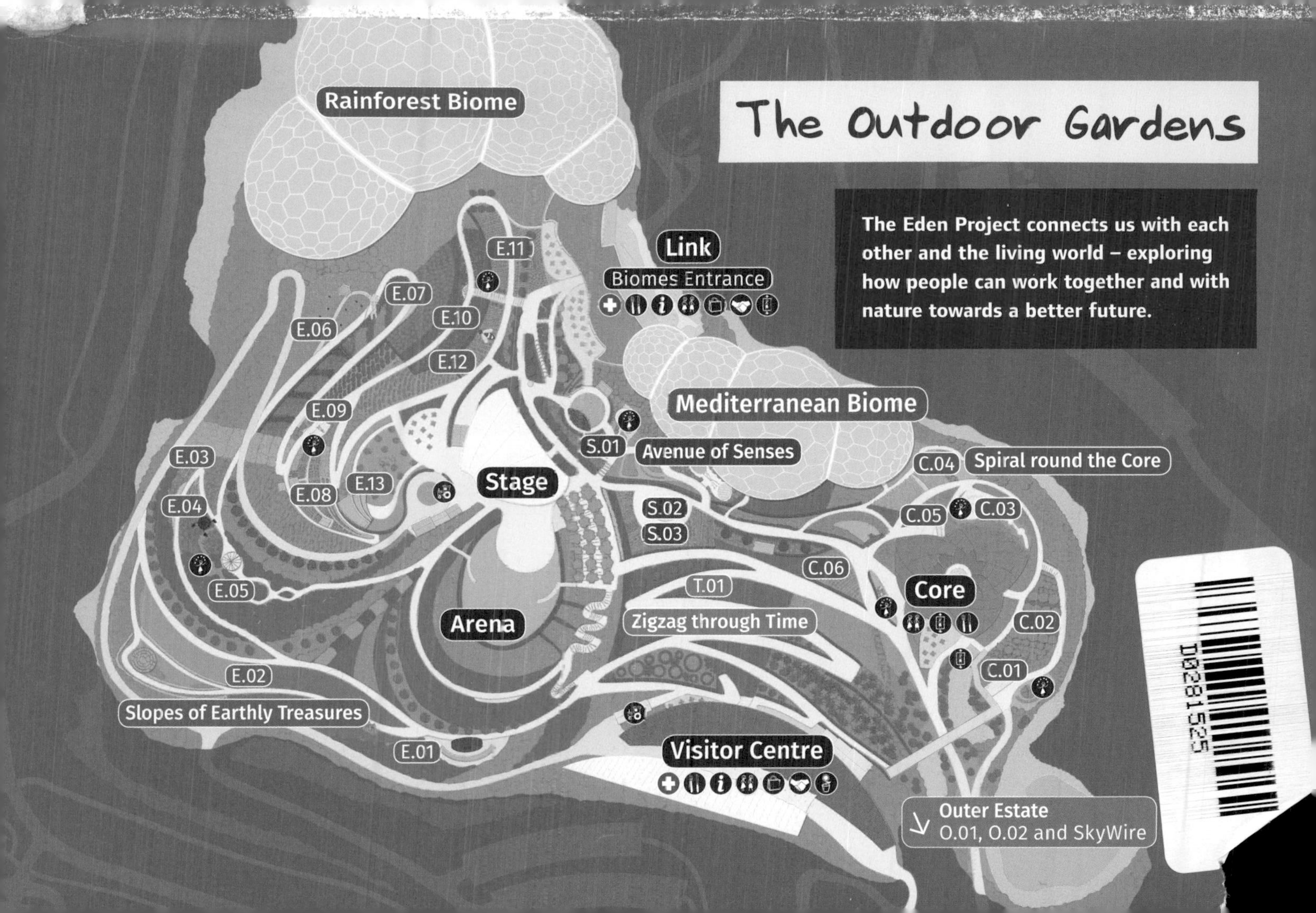

The Outdoor Gardens
The Eden Project connects us with each other and the living world – exploring how people can work together and with nature towards a better future.
Rainforest Biome
Link
Biomes Entrance
Mediterranean Biome
Avenue of Senses
S.01
S.02
S.03
Stage
Arena
E.01
E.02
E.03
E.04
E.05
E.06
E.07
E.08
E.09
E.10
E.11
E.12
E.13
Slopes of Earthly Treasures
T.01
Zigzag through Time
C.01
C.02
C.03
C.04
C.05
C.06
Spiral round the Core
Core
Visitor Centre
Outer Estate
O.01, O.02 and SkyWire

Contents

Welcome to our Living Theatre of Plants and People.

This guide gives a flavour of the sort of things we do. Come on in, explore our exhibits, take part in our events and have a chat with the team.

To keep in touch follow us online, become a Member and/or get involved in some of our projects… **edenproject.com**

Your tour guide

What is Eden about?

1999

The Eden Project connects us with each other and the living world.

The Eden Project, an educational charity and social enterprise, creates gardens, exhibitions, art, events, experiences and projects that explore how people can work together and with nature towards a better future. Our first project was to make a 35-acre global garden in a 50m-deep crater that was once a china clay pit to demonstrate regeneration and the art of the possible. In 2016 it's our fifteenth birthday. Come and join the party.

2003 - a mere four years later

The Living Theatre of Plants and People

We called our HQ the Living Theatre of Plants and People because in it we proudly present (and celebrate) our relationship with and dependence on plants: plants that feed us, clothe us, cure us, make and colour our fabrics and our lives and even supply the very air we breathe.

The Visitor Centre serves as the beginning and end of your visit. It looks out over the two vast covered Biomes which house wild landscapes, crops and stories from the Rainforest and Mediterranean regions. These bubble-shaped greenhouses are a backdrop to our Outdoor Gardens, which grow the plants from our own climate. The Link connects all three areas together and takes plant to plate with a range of delicious foods in the Eden Kitchen. The Core, with the spiky roof, pulls the threads of the story together in a range of exhibits and exhibitions and shows the work of our projects. Across the site there are many places to eat, rest and play (plus masses of loos). Take a look at the **Planning Your Day** page overleaf, the maps on the inside covers, and have a chat to any of our team, who will help you find things that would best interest you.

Why plants?

Plants are our lifeblood. The natural world is our life-support system. Each of our Biomes is divided into two main areas: the wild landscapes and the cropped areas.

Wild landscapes

Plants provide services.

Worldwide, plant ecosystems regulate our climate, provide oxygen for us to breathe, purify water, make clouds, make soil and recycle waste.

Plant leaves are the original solar panels. They capture the energy from the sun and combine it with water and carbon dioxide to make a type of sugar that all living things can then use as a source of energy and a building block for everything.

Plants also provide inspiration for design, and generally help us feel good. At Eden we bring you wild landscapes from the American prairies to the largest rainforest in captivity.

Crops

Plants provide resources. They inhabit every part of our lives including our dreams, myths and stories.

People grow crops and harvest from the wild to feed our needs. From coffee beans to blue jeans, we use plants for everything: foods, fuels, medicines, fabrics, building materials, musical instruments, sports equipment, transport, makeup, books, furniture, drinks, entertainment …
You name it, they are usually in there somewhere.

Even oil, petrol, coal and plastics have their origins in the plants of the seas and our great primeval forests.

Our exhibits explore the plant/people stories that reconnect us all to our world, to encourage us to care about it, not just because it's amazing but because it keeps us alive. It's also a shot over our bows: what we do to nature we do to ourselves. We need to crop the land and conserve the wild places.

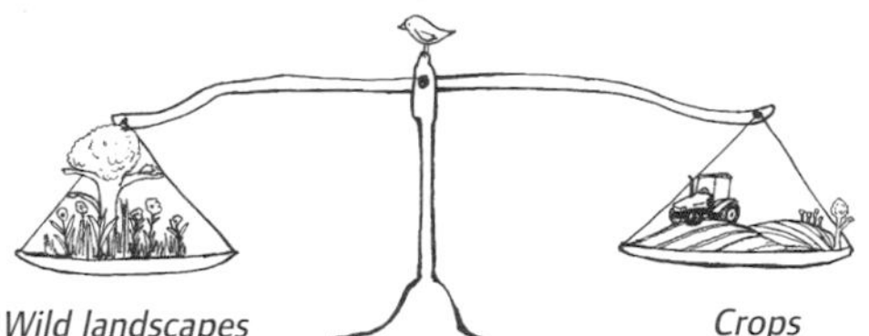

And people?

We need to live in balance with this natural world to overcome some of the challenges we have imposed on it.

As Douglas Adams (of *Hitchhiker's Guide to the Galaxy* fame) said: 'The world is big enough to look after itself. What we have to be concerned about is whether or not the world we live in will be capable of sustaining us in it.'

We are inspired by the belief that people are more than capable of changing things for the better. In the last few years we have learnt what ingenuity, resourcefulness, hope and determination can do. As time goes by we are able to show more new ways of doing things in the 21st century. You will find these stories across the site.

So, travel the world in a day: explore the exhibits, trek through the world's largest rainforest in captivity, meander through the Med, enjoy events and concerts, trace plants to plates and eat delicious foods all prepared with the planet in mind.

The Plant Takeaway – ***aka* Dead Cat, Visitor Centre.**
Discover what happens when the plants we use are taken away. Follow the adventures of Alan, Enid, Sniff the cat and Digger the dog in the children's guidebook.

Planning your day

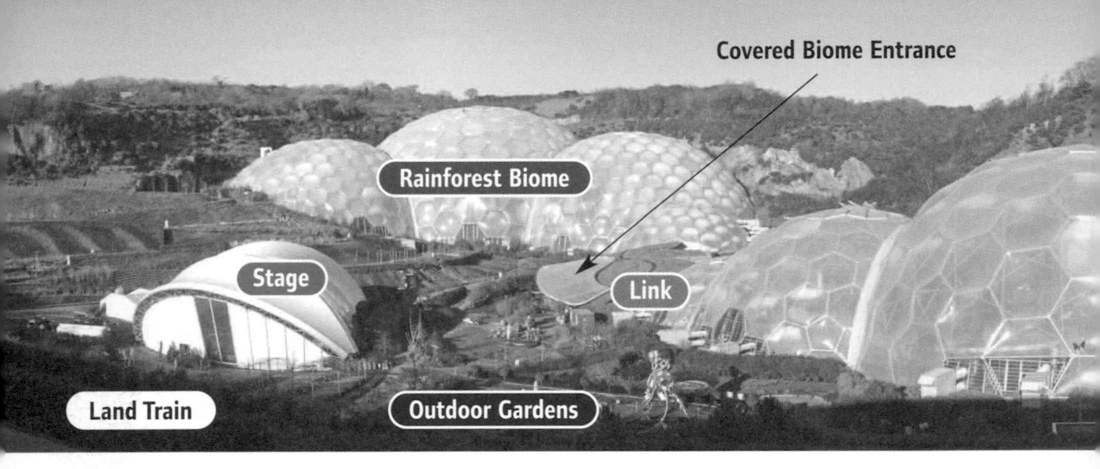

Rainforest Biome

The largest rain-forest in captivity. Trek through the steamy jungle, discover how rainforests keep us alive and how we can help do the same for them. Bananas? Coffee? Cashews? All here. Explore the need for a balance between the world's wild places and the cropped landscapes. Visit our Rainforest Canopy Walkway.

Baobab Bar open when possible.

Land Train

Every 15–20 minutes between Visitor Centre and Stage area.

Stage

The base for our seasonal events programme: The Great Easter Egg Hunt, Eden Sessions (music events), Dinosaurs, summer barbecues, Halloweden and our winter festival with ice skating and Father Christmas. For the latest programmes visit **edenproject.com**.

Great food in and around the Stage in the holidays and when there's an event on.

Outdoor Gardens

Back in 1998 this site was a barren landscape, with no soil and no plants. Today it celebrates our dependence on plants, presents our cropped and wild landscapes and explores their importance. It shows how people can work together, and with nature, towards a better future.

Link

This grass-roofed building serves as the entrance to both Biomes and the Eden Kitchen. There's also loads of loos, our little shop and Annual Pass activation station.

Eden Kitchen. Tasty food prepared and cooked in front of your eyes. Join us round our kitchen table.

Mediterranean Biome

Sights, scents and stories from the Mediterranean, South Africa and California. Visit the wild landscapes and stroll through the world's kitchen gardens with wild vines, age-old olives, the citrus grove... We need both wild and cultivated, it's all a matter of balance.

Eden Med Terrace. Authentic delicious Mediterranean food. Get in that holiday mood.

Lift and Bridge join the Core and Visitor Centre.

Core

Home to exhibitions, art, schools programmes, the 75-tonne Seed sculpture and a play area for the young ones. Discover how the planet's wild places keep us all alive and kicking. Find out about some of Eden's project work behind the scenes … and try to figure out why we built the world's largest nutcracker.

The Core Café. Open in the busy seasons for baguettes, soups, salads and hot and cold drinks.

Visitor Centre

The way in and out, tickets and Annual Pass activation station, our big shop, plant sales, ATM and loos.

The Eden Coffee House, a pit stop on arrival or departure – sandwiches, soups, cakes, coffee…

Food

The food we serve reflects the stories we tell. It is responsibly sourced: fairly traded, direct-sourced, organic, seasonal, and/or local and freshly made. We cater for vegetarians, vegans, meat-eaters and gluten-free diets. There are several cafés to choose from. Check site for opening times. Some cafés close in quieter periods.

The Outdoor Gardens

The Outdoor Gardens

Taking it from the top

Welcome to the Outdoor Gardens

Get a great overview from the platform outside the Visitor Centre and further down to the left at the Junction. In our Outdoor Gardens there are over 20 exhibits exploring the crops, gardens and wild landscapes of our own climate (list and map inside front cover). It is 15 years since we planted this once-sterile site to show that land can be brought back to life.

The entrance to the covered Biomes is in the grass-covered Link building between them.

The Zigzag through Time

The Zigzag takes you to meet the oldest plants on earth.

Route: From the Visitor Centre go left then first right at the Junction OR go right then first left. Zigzag down to the Stage and Biomes or Core depending on whether you turn left or right at the bottom. That's why we call it the Zigzag.

Prehistoric Garden T.01

These mosses, bryophytes, ferns and horsetails were around over 350 million years ago, growing in a hot steamy world way before the dinosaurs. They created coal as they rotted. Burning the coal releases the CO_2 they locked up and is heating up the planet again (more on climate change in the Core). Also look out for the planet's first flowering plants, Magnolia and Drimys, and the Wollemi pine, among the rarest and oldest plants in the world.

You can find out about the evolution of Eden over the past 15 years on pages 42–47.

The Slopes of Earthly Treasures

Enjoy a range of stunning gardens on the Slopes of Earthly Treasures, growing plants that feed us, clothe us and keep us warm, plants for gardens and tricky locations, plants in stories and in the wild. Tales of rocks too – everything we use is either grown or mined.

Route: From the Visitor Centre go left past the land train station then up to the giant tyre. There are great views of the site from up here and some wonderful hidden gardens. Alternatively take the land train and get an overview as you travel down. And/or you can get the land train back up. Choices, choices.

Mining the earth: metals, minerals, energy E.01

We're as dependent on mining as we are on agriculture. Eden is a great example of how an old mine site can be reclaimed when the mine has reached the end of its useful life. The giant wheel tells the story of the copper on the Core roof: responsibly sourced from a single mine. Climb into the mini tunnel to discover underground secrets and how the industry is working towards good practice.

Wild Cornwall E.02

Cornwall's biodiversity, its varied landscapes, habitats and wildlife, has been shaped by its climate, geology, geography and people. Heathlands are partly man-made; they started forming 6,000 to 3,500 years ago when woodland was cleared for hunting and agriculture, and need managing to prevent them reverting. Since 1800, Cornwall has lost over 90% of its lowland heath. Conservationists help to restore, protect and manage what is left. In Cornwall 74% of the land is farmed, though farming also provides rich habitats. Here, sculptors Peter Martin and Sarah Stewart-Smith immortalised rare Cornish species in stone, and Chris Drury created the Cloud Chamber.

Plants for a changing climate E.03

Our extreme gardeners explore adaptation and try new plants in tricky situations.

Biomass fuels E.04

David Kemp's 'Industrial Plant' sculpture takes a sideways look at fossil fuels, which provide around 87% of the world's energy. The developed world (15% of the world's population) uses 53% of this energy, while nearly half the world's people (mainly in developing countries) rely on wood, charcoal and dung, which have a low carbon footprint; the amount of CO_2 absorbed during their growth equals the amount emitted on burning. Willow, poplar and miscanthus are burnt for biomass in the UK. Other energy choices include wind and water power, algae as fuel, solar and nuclear power and geothermal, using heat from the earth.

Myth and folklore E.05

Stories keep plants alive in our memory, and our storytellers bring you tales across the site. Pete Hill and Kate Munro created our willow Dream Chamber. Its classic seven-ring labyrinth is found worldwide. To sailors it was a good-luck token, ensuring safe return. It provided protection against wandering spirits who get lost in the curves (spirits can only travel in straight lines, allegedly).

Prairie E.06

Looks great in full flower in August. The American prairies were partially created by man, using controlled burning to attract game to young post-fire grass. They once covered a quarter of the US. In some areas, up to 99% have been destroyed in the last 150 years. Work is underway to conserve these diverse grasslands and let the buffalo roam once more. Why conserve? Provision of ecosystem services (see the Core, p.20), climate control and potential future crops for starters. We manage our prairie by burning in Jan/Feb, then the plants start to come through: Camassias, Liatris, Echinaceas ... Some public parks are now turning to prairie-style plantings: better for biodiversity, cheaper and easier to maintain than bedding.

The Slopes of Earthly Treasures (continued)

Plants for materials, plants from Cornwall, plants for bees, plants for supper, plants for thirst-quenching teas and tasty beers, cheers.

Crops for a material world E.07

Grow fibres, fuels and ... plastics. Industrial hemp provides food and health products, clothing, car components and building materials and has a lower carbon footprint than concrete. It grows well in the UK, needing few agro-chemicals (cotton uses around 25% of the world's pesticides). Legally we have to have a barrier round our hemp crop, so George Fairhurst designed us a hemp fence. He made our Metal Giant too. Give the rope a tug to see how strong plant fibres are. Sunflowers provide high-quality oils for lubrication, plastic manufacture and biodiesel as well as food. Plant sugars and starch from maize and wheat make bioethanol for fuel and compostable plastic cutlery, carrier bags, nappies, etc. Plants grown for biofuels are having their carbon footprint scrutinised, and the food-versus-fuel debate means a search for alternative raw materials. Biomass to Liquids (BTL) factories can turn waste straw, wood waste, stalks, etc, into fuel, even biokerosene to power aircraft.

Global Gardens E.08

Discover new veg to grow in your garden, from gardening communities with roots in Africa, Asia, Latin America, the Caribbean and Europe. Find out more about Eden's Gardens for Life project: a global network of schools that explore the world through gardening and growing food (p.55).

Tea E.09

Made from the young leaves of the tea bush, *Camellia sinensis,* tea grows in the subtropics, as well as the cool, moist, more mountainous tropics – and in Cornwall.

Beer and brewing E.10

Look at our hop poles for beer ingredients (barley and hops), the hop stilt walker, the brewing kettle and the isinglass (tropical catfish swim bladder) used to clear beer.

Cornish crops E.11

Cornwall's mild climate means that it can supply the UK with early crops and quality foods year round. Eden's catering uses local produce wherever possible; 91% is bought in Cornwall and Devon.

Pollination E.12

Plants can't move (much). Many reproduce by luring insects and other animals to take their pollen from flower to flower. Insect/flower relationships are often very specific. Over a third of our food plants worldwide depend on pollinators – let's look after them. Lavender, from the Latin *lavare* ('to wash'), cleans and calms. It is used in aromatherapy oils, perfumes, insect repellents and antiseptics. Bees love it too and make delicious honey from it.

The Hive in the Orchard E.13

In our Hive building enjoy seasonal workshops (occasionally closed for private functions). Orchards: a place of relaxation and, in autumn, fast food (in a biodegradable wrapper). Only around 12% of our annual fruit consumption is produced in the UK, but interest in health and local produce means that orchards are making a comeback. Try growing some tasty old varieties, start a community orchard near you, have an apple day. Try some Cornish apple juice in our shop.

The Avenue of Senses

Here you'll find gardens for all the senses and the beginnings of a really important garden which will tell the story of the crops that feed the world and the water that nourishes them and us all. Come back and watch this one grow.

Route: These exhibits are all on the main path between the Core and the Link Building (the entrance to both covered Biomes).

The Garden of Senses S.01

There are around 500,000 hectares of gardens in the UK (twice the area of nature reserves). Our own gardens represent a personal connection to nature, serving all sorts of purposes.

A Sense of Taste A productive garden half this size can provide a family of four with fresh veg all year round. If you are short of space just grow your favourites. Rocket takes only five to six weeks from sowing to eating.

A Sense of Scents Herbs, most of which don't need much water, are perfect for growing in pots. You know it makes scents.

A Sense of Pukka Pukka means 'real, authentic or genuine'. We're working with Pukka Herbs to tell stories of how plants look after us and how we can look after them. Discover Grow Wild (responsible wild harvesting).

A Sense of Place This social garden has the environment close to its heart. Drought-resistant plants were chosen to avoid watering, and paths made from local materials to avoid air miles.

A Sense of Play Our play garden is made of willow and wood, helping to reconnect children with nature.

A Sense of History Garden flowers have their origins in wild places across the world. Favourites were bred to produce the garden varieties we are familiar with today. The dahlia originated in Central America; stems of its ancestors were once used as water pipes.

A Sense of Memory Send your fond memories of particular plants to **memories@edenproject.com** and we'll put what we can online. This garden, created by Thomas Hoblyn for the RHS Chelsea Flower Show 2011 with Homebase, was given a permanent home at Eden.

A Sense of Colour A range of borders with themed colours: red, white and blue create a stunning impact. Planting schemes are available from our plant shop and **edenproject.com.**

WEEEman S.02

Our waste giant is made from all the Waste Electrical and Electronic Equipment (about 3.3 tonnes) one person throws away in a lifetime. Designed by Paul Bonomini.

Crops that feed the world S.03

Three plant species – wheat, maize and rice – feed most of the world. They provide carbohydrates and some protein, and store well. Other major staples include potatoes, beans and bananas. Rapid world population growth led to wheat and rice breeding programmes in the 1960s and 70s to increase crop yields. This 'Green Revolution' staved off predicted large-scale starvation.

Wheat for...

Bread
Cakes, biscuits
Pasta
Wheat noodles
Animal food

Maize for...

Corn on the cob
Cornflakes
Tortillas
Animal food
Popcorn

Rice for...

Rice (steamed, fried, etc)
Rice Krispies
Risotto
Rice cakes
Rice pudding

Today's challenges include a growing population (over 7 billion, possibly growing to 10.5 billion by 2050); inequity in distribution; over-consumption and cheap food in many developed countries and hunger in many developing countries; a lack of reserves; rising oil prices; the food/fuel debate (see exhibit E.04); land grabs; price hikes and climate change. Possible solutions include increasing yields and soil fertility, pest and disease resistance, biotechnology, agricultural biodiversity, new crops that can cope with the changing climate, trade justice, conflict resolution, poverty alleviation, good nutrition projects and education. On a personal level, think about what we eat and where it comes from, waste less, eat less meat (63% of maize production is used as animal feed) and grow your own.

The Spiral round the Core

Play, playful and peaceful gardens with plants for timber, dyes, paper and health. There are secret routes through wooded slopes and a great tranquil spot out the back too.

Route: From the Visitor Centre go right and across the Bluff Bridge and down the lift or the steps to the Core and its gardens.

The Bluff C.01

A place for picnics and play for children of all ages.

Timber C.02

The plant labels are made from the timber they describe. Wood is a carbon store and timber construction is on the rise. Good for wood.

Spiral Garden C.03

Making a school or community garden? Discover some low-cost ideas including willow spirals, rainbows of flowers, soft paths, textured plants and scented plants.

Paper and play C.04

Plants grown here can be made into paper. Many natural habitats have been cleared and planted with Eucalyptus and *Pinus radiata* for paper-making. Agricultural residues containing cellulose, such as wheat and rice straw, can be used instead of wood pulp. World paper consumption is rocketing – so is recycling, luckily. Look out for the secret paths through the tall grasses and the log house to play in.

Dyes C.05

Woad and indigo give us blue, weld yellow and madder red. Indigo-dyed cloth comes out of the dye vat yellow and turns blue as it oxidises in the air.

Health C.06

Half the herbs sold worldwide are wild-harvested, which done responsibly can sustain environments and livelihoods. More on this in exhibit S.01. Herbs and pharmaceuticals (e.g. morphine from opium poppies) are also grown as crops. With 'pharming', crops are modified and used as biological factories to produce drugs: pills from plants.

Hangloose

SkyWire Fly 660m at up to 60mph, on the longest and fastest zip wire in England. Fly upright or headfirst in our 'superman' harness.

The Gravity Giant Swing New extreme thrill ride. You'll be suspended 20m high then plummet in a breath-taking 50mph free-fall back down towards the earth. Book online or at the Hangloose shed in Pineapple car park.
hangloose@adventure.com

The Outer Estate

Wild Chile 0.01

Behind Pineapple car park take a trip to Chile and explore our forest of beautiful plants. Central Chile is under threat from logging, agriculture and replacement with non-native pine and eucalyptus for paper and wood chip. This 'Safe Site' contains and protects wild-collected material from the Valdivian forests of central Chile, like this *Embothrium* – a living example of *ex-situ* conservation.

Forest Garden 0.02

On your way down to the Visitor Centre or back to the car parks, explore our growing Forest Garden – full of useful wild plants for food, shelter and medicines.

The Core

The Core

Eden's education, arts and events hub

The building's structure is based on a sunflower, which isn't one flower but hundreds that combine to create something better. Eden's projects, many described in the Core, also show what can be achieved by working together. The Seed, a 75-tonne Cornish granite sculpture carved by Peter Randall Page, carries nature's pattern into the heart of the building. It was inspired by the growth blueprint of plants, opposing spirals based on Fibonacci's sequence: 0, 1, 1, 2, 3, 5, 8, 13, 21, where every number is the sum of the previous two. The spirals on a pineapple, a sunflower and our roof exhibit two consecutive numbers in this sequence.

The Core building

The design Built with responsibly sourced materials and energy efficiency in mind. The timber structure comprises Swiss Forestry Stewardship Council-endorsed double-curved glulam (glued laminated) beams. The floors consist of recycled wood, Marmoleum (made from flax), carpets (made from maize), concrete from china-clay sand (low carbon footprint) and green tiles from recycled Heineken bottles. The Core is on three floors and built into the landscape so that each floor is accessible from ground level. It was designed by Grimshaws and funded by the Millennium Commission, the South West Regional Development Agency and Cornwall's Objective One Programme.

The roof 40% of the world's copper is recycled. What about the other 60%, we wondered? Often metals mined from different places are mixed together so it is difficult to ascertain their source. Here we worked with the industry to trace the supply chain of our roof's copper from a single Rio Tinto mine, known for its high social and environmental standards. This unusual initiative has led to more work on mineral supply chains.

The Core exhibitions, installations and exhibits

Ground Floor The huge glass ball, the Plant Engine, represents the world's ecosystems: rainforests, oceans, grasslands, etc. It breathes life into the bell jars containing automata which show the 'ecosystem services' that keep us alive: regulating our climate, providing oxygen for us to breathe, purifying water, providing inspiration for design … Challenges humans impose on these systems – climate change, clean water provision, biodiversity loss – are explored in the greenhouse, water tank and Diversity Cabinet, along with some solutions.

Carts representing Eden's projects for positive change can be found in the Core and on p.54.

Biodiversity *= Life in all its variety. We're part of nature and it keeps us alive. Human impact is causing many species to disappear, and with them patterns of life and processes that our societies depend on. Biodiversity loss is an issue. If it goes too far we will be among its first victims. Check out the Diversity Cabinet for loads of stories.*

Play We have recently installed a play area in the Core. Eden loves play: it encourages investigation, curiosity and learning, helps a reconnection with nature, has social elements and provides (managed) risk – all vital for growing rounded citizens.

First floor

New exhibition for 2015
Invisible You – The Human Microbiome
We are not alone. Bacterial cells outnumber our own 10:1. These 'communities', within and on our bodies, are essential in the main. They help to form, feed and protect us. They help our immune systems develop, help digest the food in our gut and some of them protect us from the bad bacteria that make us ill. The exhibition's story is told through a series of stunning art installations.

Funded by the Wellcome Trust.

Classrooms: for our schools programmes and events

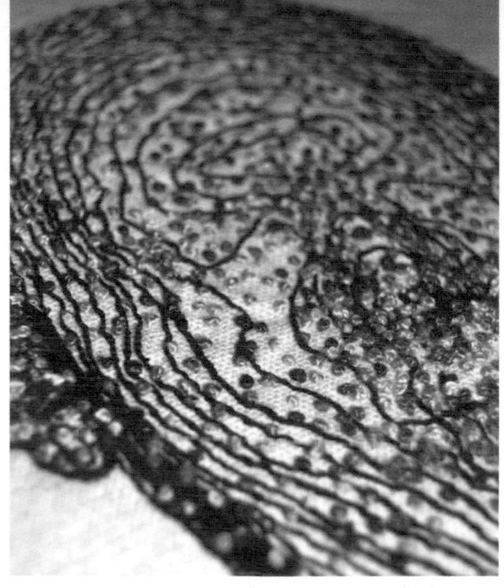

Invisible You tapestry. Rebecca Harris

Second floor

In our solar terrace windows don't miss the beautiful array of photograms created by Susan Derges that represent the water cycle.

The Core Café: delicious food including baguettes, soups and salads (closed on quieter days, but plenty to eat in the Eden Kitchen in the Link between the covered Biomes).

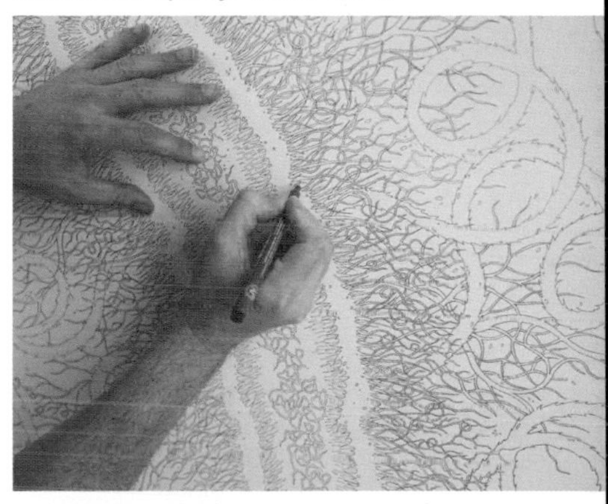

Invisible You papercut (in progress). Rogan Brown

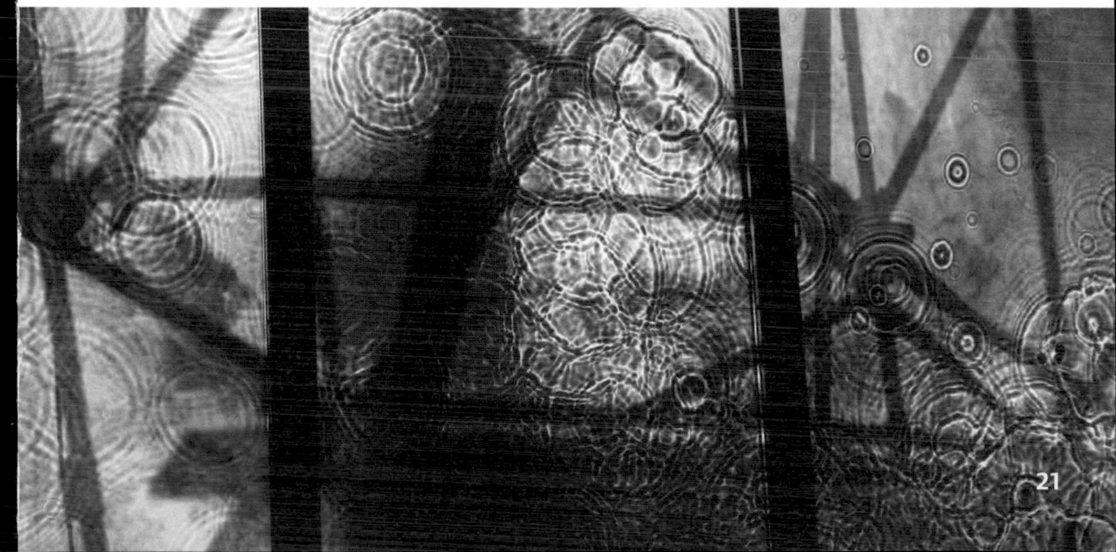

Photograms. Susan Derges

The Rainforest Biome

'Rainforest is the glue that holds the climate of our planet together. Lose the forest and it will have devastating consequences for all life on Earth.'

Professor Sir Ghillean Prance, FRS

The Rainforest Biome

Welcome to the Rainforest Biome

The Humid Tropic Regions are located between the Tropics of Cancer and Capricorn, 23.5° N and S of the Equator. Temperatures average 25°C all year round (5°C variation), with over 90% humidity and 1,500mm (60″) annual rainfall. We've recreated these conditions here (except we do the watering before you arrive) and in the last 15 years have grown the largest rainforest in captivity. Why?

The majestic rainforests that circle the globe are the planet's life-support system. We can't all go there, so we decided to grow some here. We explore their secrets, discover how they keep us alive, learn of the challenges and, most important, share ways of trying to save them. On your Eden journey you can trek through four of the world's rainforests. The plants in each area may be different species but they have evolved to look similar to thrive in the climatic conditions. The first phase of our Rainforest Canopy Walkway takes you into the treetops to explore biodiversity close up.

Millions of people live in and make their living from the rainforests; we bring you some of their stories. Half the Biome grows crops: chocolate, coffee, spices, rubber, medicines... Traditionally forest was felled to cultivate some of these things. We explore the search for balance: finding ways to grow crops to meet people's needs (supporting livelihoods as well as providing consumer goods) whilst conserving the forest that keeps us all alive.

How does the rainforest keep us alive?

- Rainforests help regulate the earth's climate. They absorb and store CO_2 in their wood. They make a lot of rain too.
- They are a huge store of biodiversity. They cover around 5% of the earth's land mass and are home to over half the world's terrestrial animal and plant species. Some have been used to keep us fed, clothed and cured. Around 99% of these species have yet to be studied.
- The forests produce oxygen (but not as much as do the algae in the sea).

An area of primary forest the size of this Biome is cut down every 10 seconds. Don't panic (yet); rainforests can regrow. There are many things we can do too – read on.

Eden worked with Survival International to produce a thought-provoking photo exhibition of tribal peoples of the rainforest by the renowned photojournalist Sebastião Salgado and Cornish explorer and writer Robin Hanbury-Tenison, shown here with one of his photographs.

Explore the world's rainforests

Tropical islands: conserving the land R.01

Mangrove swamps protect the coast (e.g. against tsunamis), provide fuel, timber and are an important habitat for fish, also acting as a 'nursery' for most coral reef fish. Island habitats are home to many unique species. Climate change, invasive species, human settlement and tourism pose serious threats. Isolated island communities lack resources to conserve biodiversity, though conservation programmes offer possibilities. The rare Seychelles Coco-de-Mer, with huge seeds that look like giant bottoms, was over-harvested (as trophies and perceived aphrodisiacs). Now each seed is registered and protected. Ours (a gift from the Seychelles) germinated five years after planting: one of very few specimens in the UK, it grows slowly – a leaf a year.

Southeast Asia: Orang dan Kebun (people and garden) R.02

Our traditional Malaysian home garden provides food year-round. Herbs and flowers nearest the house, vegetables, fruit and other useful trees further out. Winged beans replace our runner beans; both fertilise the soil. Pak choi, taro and rice replace cabbage, carrots and potatoes.

The garden also provides building materials, medicines and produce to barter or sell at local markets. The miracle tree, *Moringa oleifera*, has edible leaves, beans, flowers and roots. Its seeds are used as water filters and its oils for watchmaking.

To the left of the path is a rice paddy. In Asia 'rice is life', culturally and spiritually crucial to people's lives.

West Africa: managing the land R.03

The totems, by West African sculptor El Anatsui, came from charred timbers recycled from a part of Falmouth docks which was destroyed by fire. They started their life as trees in West Africa.

On your right you will find the Trees of Life exhibit. West African rainforests have over 3000 species of fruit trees. These little-known fruits and nuts are important resources for everyday life; widely consumed and locally marketed. Work to harness their potential to be cultivated as new crops is now underway.

Further round, the chop farm, where areas of forest are cleared, grows light-loving crops such as groundnuts, cassava, rice, millet and traditional African leafy vegetables alongside pawpaw and bananas. Reverting to traditional crops in these areas rather than high-status Western crops helps provide a balanced diet and an income from local markets.

Tropical South America: shamanic art and cassava production R.05

Past the Rainforest Canopy Walkway entrance (described overleaf) take the steep and stepped high road past the waterfall for a great panoramic view and see the work of Peruvian shamanic artists Montes Shuna and Panduro Baneo, showing a spiritual connection between plants and people. The flat low road takes you past the tallest tree in the Biome (the kapok).

The paths meet near a clearing where a hut shows the processing of cassava into tapioca. Cassava varieties contain prussic acid (hydrocyanide), which has to be removed before cooking

Rainforest Canopy Walkway R.04

Take to the trees. Our Walkway is fully accessible to wheelchairs and pushchairs.

People and the forest The Baka people from the rainforests of the Cameroon, the Congo basin and Gabon are deeply connected to nature. The forest is their mother, father and guardian, providing their food, medicines and shelter. They have hunted and gathered, led a very low-impact life and sung and played their way through life for thousands of years. You can listen to their music in the base-camp. Today their way of life, superb listening skills, deep forest knowledge and the forest itself are under threat.

Up on the Walkway, discover what canopy scientists get up to and spot their IKOS pod – an authentic aerial campsite.

> 'Rainforest peoples are the original tropical ecologists. We still have much to learn from them if we are to manage rainforest ecosystems in a sustainable manner. I have often said that I have learnt more from Amazonian tribal elders than I ever did from teachers at university. The recording and preservation of this indigenous knowledge is vital to our future survival.'
>
> Professor Sir Ghillean Prance, FRS

Funding and phasing

Thank you to the Garfield Weston Foundation, the Wolfson Foundation, donors to the Eddie George Memorial Appeal and donations from other organisations, individuals, visitors and Friends who made this first phase possible. We are currently fundraising for the next phase, where we will make weather and explore climate. **edenproject.com/walkway**

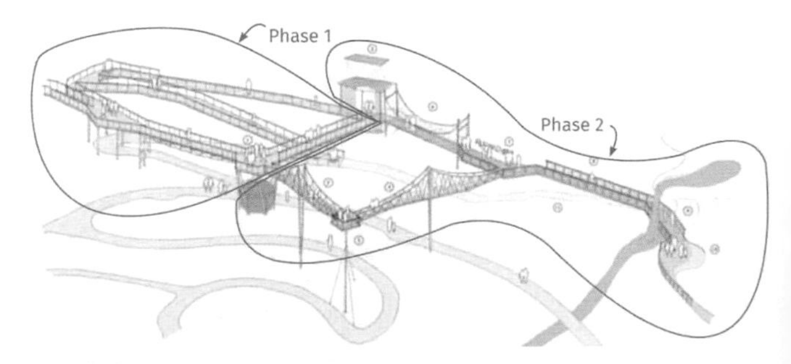

'The Rainforest conservation projects which are successful are those which work with the local indigenous people.' Andrew Mitchell, globalcanopy.org

Plants and the forest Look out for labels carrying the stories about the tremendous trees as you walk towards the Nest platform.

The Nest platform Plants eat, drink, reproduce and protect themselves from danger – without moving. Our Biodiversity Chandelier explores nature's designs which have produced forms that fit their function: from fiercely protective prickles to extraordinary pollination mechanisms

Challenges Forests are cleared for agriculture, mining, development and timber. 12–20% of carbon emissions (which contribute to climate change) come from deforestation. Over 100 rainforest species may be lost every day. To ensure our future survival we need to help the forests to survive.

Solutions Rainforests can regrow or be replanted and managed sustainably for the future.

- **Share** what you discover about how forests keep us alive
- **Support** charities and organisations working to save the forest
- **Shop** for products which look after the forest (eg. FSC, Rainforest Alliance) and avoid ones that don't (not just oil palm; forests are often felled to grow soya and meat)
- **Write** letters on rainforest issues to politicians
- **Volunteer** for a rainforest charity

Welcome to crops and cultivation R.06

Beyond the arch, find the plants that produce things we use: rubber, cocoa, chocolate, bananas. It's all a matter of balance between conserving the wild and cropping the land.

Growing with the forest R.07

'Slash and burn' is a common practice used by desperate peasant farmers. Alley-cropping, where crops are grown between rows of Inga trees, is a new alternative. This agro-forestry system is good for the forest and the people. The trees, like their relations peas and beans, contain special root bacteria which make nitrogen fertiliser from the air. The trees also provide shade, fruits (ice-cream bean) and can be cut back and used as a weed-suppressing mulch which rots down to make compost. (NB: Not all legumes – pea and bean trees – have edible fruits; some are poisonous.)

Regrowing the forest R.08

Rainforests can regrow. Replanting pioneer trees quickly creates canopy cover, suppresses weeds, attracts animals that bring in seeds and nurtures timber tree species. Eden works on and with projects that help.

Rubber R.09

Rubber trees, *Hevea brasiliensis*, grow wild in the South American Amazon rainforest and have been tapped for rubber for boots and balls for centuries. Growing demand for car tyres in the 20th century stimulated the plantation industry in Asia, where cultivated trees replaced rainforest. Demand for natural rubber is still high. Condoms, rubber gloves, shoe soles and many other products still use natural rubber.

How can we help save the rainforest, harvest important resources and enable the rainforest peoples to make a decent living?

WWF and Sky Rainforest Rescue worked with the Acre State Government in Brazil to help save one billion trees and make the forest worth more alive than dead. Find out about one aspect of the project, working with local rubber-tapping communities to help make rubber tapping economically viable for local farmers, here at Eden.

Photo: John Lennon

Ghanaian cocoa farmers from the Kuapa Kokoo Co-operative admiring Eden's cocoa crop

Cocoa and chocolate R.10

Cocoa originated in South America. Today it is grown by around 4.5 million farmers, mainly on West African smallholdings. The UK chocolate industry supports schemes like Fairtrade.

The Mayeaux Tapestry in the Rainforest Biome takes you through the chocolate timeline.

The Latin name for cocoa, *Theobroma*, means 'food of the gods'. Our chocolate 'extra-ordinary' ice-cream is made with fairly traded 'fino de aroma' Colombian cocoa: top-notch, said to be the best in the world.

Scientists are crossing different wild species (from South and Central America) with improved varieties of cocoa trees to create disease-resistant trees which means growers can use fewer chemicals and plant less on new land.

Stories from the rainforest R.11

Looks at what you can do to make a difference (see p.27).

Palms R.13

Many palms are used in the tropics, for food, fuel, walls, thatch, ropes, boats, sago, sugar, cooking oil... However, one palm has become a global commodity. Oil palm produces 'vegetable oil', found in nearly half of bestselling supermarket lines, even cleaning products and cosmetics. Orangutans' habitats are felled for oil palm plantations. We don't use palm oil in our Cornish pasties (yes, even pasties sometimes contain palm oil). Global supply chases demand, and plantations march into the rainforest. The Roundtable on Sustainable Palm Oil works on sustainable production. **rspo.org**

Tropical fruits – Bananas and friends R.13

Bananas Over 85% of the bananas grown in the tropics stay there, a staple diet for millions. Different varieties provide savoury or sweet dishes, juice, wine and beer. Exported bananas are generally Cavendish types from large plantations (such as those in Latin America, usually owned by large companies) or from smallholdings (such as those in the Caribbean, usually owned by local farmers). Organic and/or fairly traded bananas are available in shops: your wallet is your weapon. Panama disease has blighted some Cavendish crops worldwide, so growers are trying new varieties – and so are we (with our 'Formosana', a resistant Cavendish variety).

Mangoes For us a delicious treat, for some a promising export crop for developing countries and for others a vital famine food.

Pineapples No, they don't grow on trees. Fairtrade and organic pineapples are on the increase.

Sugar R.14

Sugar is made from tropical sugar cane (and temperate sugar beet). In the 1300s we each used around a teaspoon a year. Today it's around 30kg, and stories of diabetes, obesity and poor teeth abound. Sugar cane is also used for ethanol production. According to the industry, this doesn't get caught in the food or fuel debate (unlike maize) and doesn't displace rainforest. Bagasse, the waste product, can be burnt to generate electricity. Over 7m people are employed worldwide in the sugar industry, many in developing countries. Fairtrade and organic sugar are on the up; the sugar used at Eden is fairly traded. We also use and sell traditionally and sustainably produced panela, a type of natural cane sugar, from Colombia – it has a wonderful toffee taste. Try our panela ice-cream.

Rice R.15

One of the most widely grown grasses in the world. Farmers grow it to eat and sell the seed, some to us. Scientists are working to increase yields to feed the growing population and are also working in partnership with poor farmers, exploring how to grow rice in different and difficult environments.

Tropical fruits – Baobab and friends R.16

We serve refreshing baobab smoothies here when possible. African baobab is called the tree of life. It can provide shelter, clothing and water as well as food.

Our baobab fruit comes through Phyto-Trade Africa. Money raised supports the harvesters. Our baobab bar is surrounded by other tropical fruits; soursop, rambutan and the unusual jaboticaba tree.

Bamboo R.17

Bamboo is used by half the world's people to make homes, furniture, food, fuel, music, medicine, scaffolding and suspension bridges. Its hollow tubes are strong but light. Within its tissues short, tough fibres sit in a resilient matrix: nature's fibreglass.

Coffee R.18

Coffee is big business. The beans *(right)* grow inside coffee cherries which ripen at different times, making picking tricky. Eden uses Direct Trade coffee grown on specific farms in Minas Gerais, Brazil. The farmers Ricardo Barbosa & Virgulino Muniz care for their workers and the environment. It is sourced for us – direct – by the Cornish company Origin, meaning fewer links in the coffee journey chain. They roast it in the most energy-efficient roasters too. Keep your eye out for Direct Trade. It's happening in coffee – where next?

Nuts and spices R.19

Today spices are cheap. In the past they were worth their weight in gold and shaped the world as we know it. Explore the stories on the spice boat and see if you can solve its riddles.

Nutmeg Though nutmeg was thought to cure the bubonic plague, it was along the spice route from Central Asia that the Black Death first travelled to Europe in the mid-1300s.

Ginger Has many relations: turmeric, cardamom, torch ginger, galangal, used for flavourings, starch production, antiseptics and in cultural ceremonies. Some gingers are NTFPs (non-timber forest products), harvested from the rainforest without harming the trees or environment. Good for plants, good for people.

Cashews Why are cashews so expensive? Roasting, shelling and cleaning the nuts is laborious and the shells contain highly corrosive cashew-nut shell liquor (CNSL). Traditionally used to treat ringworm, CNSL is now sometimes used in heatproof enamels.

Secrets of the rainforest R.20

Come and enjoy the beauty of canopy plants like these bromeliads.

The Mediterranean Biome

The Mediterranean Biome

Salutare! Enter all – The Lands of the Warm Temperate Regions
Journey through:
The Mediterranean's Paradise
South Africa's Garden
California's Horn of Plenty
Born of Sister Fire and Brother Drought

Immerse yourself in Culture's Cradle
Taste history, olives, citrus and wines
In the world's kitchen gardens
Born of Water and Mankind

Welcome to the Mediterranean Biome

Mediterranean-type climates with their hot, dry summers and cool, wet winters are located 30–45°N or S latitude on the western sides of continents. They are within the Warm Temperate climate zone.

In this Biome we invite you to meander through the wild landscapes of the Mediterranean, South Africa and California. Wild? Through history many of these landscapes have been shaped by human cultures.

The native plants in all these regions have to cope with drought and poor, thin soils. Some have small, grey, hairy leaves, some make protective oils, some are spiny, evergreen and/or waxy. This may help to reduce water loss and make the plants less appetising to predators.

The plants are tough but their environments are fragile: intensive grazing erodes soils, imported plants threaten native species, land is developed and precious water is used to serve the needs of crops and people, many of whom are occasional visitors. Big pressure on small lands.

Crops? The addition of water and fertilisers to these sunny regions has created massive kitchen gardens for vegetables, vines, fruits and flowers. Take a look around. You'll find many familiar products come from these places: wine, olive oil, perfumes, lemonade.

Through the gates to the left, you will also find a typical Mediterranean courtyard garden.

On your journey, look out for Dionysus, the great bronze bull. He stands centre field, straddled between the wild and the cultivated lands. As he discovered (to his peril), it's all a matter of balance. It's in our power to leave the world better than we found it.

Perfume Garden M.01 *Coming soon*

Try to describe a scent without referring to another smell. Tricky, isn't it? The scent of violets, a whiff of mint – scent goes straight to the seat of emotion and memory in the ancestral core of your brain. Plants use scent to attract pollinators and repel predators. Do we use it to signal, seduce or warn, like plants, or for sweet memory and comfort?

The Mediterranean Basin M.02

On the lower route (no steps), the 'Liquid Gold' mosaic path by Elaine Goodwin celebrates the tradition of olive oil as a symbol of life and divinity. Look for the subtle dove images – one for each nation in the Mediterranean.

The Mediterranean landscape is mainly man-made, cleared for crops over many thousands of years, including the olives and vines that helped shape this region's civilisation and biodiversity. People leave their mountain farms seeking work on the coast, although some city dwellers are returning to smallholdings. Buying traditional foods and natural products, seeking out quality and taste, farm holidays: all can help conserve these fragile environments and communities.

Past the bell tower, the upper stepped route takes you through Maquis and Garrigue to the viewing point.

The French underground movement in World War II was called the Maquis because they hid in this hilly landscape of prickly oak, juniper and broom, a habitat that contains unique plants, insects and reptiles, but can get overlooked, having no spectacular birds or mammals.

South Africa M.03

The Cape Floral Kingdom has over 1,400 rare or endangered plant species. It covers less than 0.5% of the area of Africa but is home to nearly 20% of the continent's flora. Fynbos, with around 7,000 species (5,000 unique to this area), covers 80% of this kingdom.

'Fain-boss', Afrikaans for 'fine bush', refers to the evergreen, fire-prone shrubs in this nutrient-poor soil. Plants include rush-like restioids, shrubby heathers and proteoids, with their stunning feather-like blooms. Formed millions of years ago from the ashes of drought-stressed forests, the Fynbos has been fire-managed for conservation since the 1960s; the seeds of the Protea *(right)* need exposure to smoke to germinate. The Fynbos is threatened by agricultural and urban development, uncontrolled fire and invasive alien tree species.

Little Karoo In this semi-arid valley behind the southern coastal mountain range, the land bakes to 50°C in summer, freezes in winter and suffers severe droughts. Today much of the valley is irrigated for crops, but the surrounding hills house ice plants, aloes and types of daisy.

California M.04

California has vast landscapes and a huge diversity of plants. Ceanothus and Californian poppies, familiar garden plants, grow wild here. The spiky chaparral, with its less familiar buck bush, toyon and scrub oak (the 'chaparro'), gave its name to chaparreros, or 'chaps', worn by cowboys to protect their legs. Chaparral, grassland and the open oak forests were the results of thousands of years of controlled burning by Native Americans. California once had a rich natural harvest that was cultivated gently and sustainably by the indigenous people. Today water is so valuable that environmental activists take out court orders to make farmers leave minimum flows in rivers.

The state's high consumption brings social and environmental costs, putting it at the forefront of climate change. On the other hand, the region is also the birthplace of innovative new technology and home to some of today's most environmentally conscious people. The catalytic converter started life here and 2010 saw the first legislation to give polluting companies such as utilities and refineries financial incentives to emit fewer greenhouse gases.

Welcome to crops and cultivation M.05

Many of our crop exhibits are seasonal, so please expect changes from time to time. Producing crops in the Mediterranean regions is an intensive industry using fertiliser, water, sprays and often immigrant labour. Pressure is mounting to move to low-input, energy-efficient, diversified farming. Water is becoming increasingly scarce; many aquifers have more water extracted from them than they receive. Technologies are being developed to reduce usage and explore creating fresh water from sea water and waste water.

Cork M.06

Agriculture and conservation work hand in hand in the bio-diverse cork oak wood pastures. Cork bark is harvested and Iberian pigs, which feed on the acorns, help maintain the plant diversity and provide high-value ham. These trees, unlike others, don't die when their bark is cut off, so buying cork products supports these environments. Currently supply is outstripping demand. Organisations such as WWF, the RSPB and the Portuguese Cork Association work to conserve these wood pastures. Buying wine (and champagne) with real corks helps, but cork has many other properties and uses. It can cope with extreme temperatures, so can be used as engine gaskets, for example. Maybe new industrial applications could help save these landscapes?

Seasonal crops M.07/M.08/M.09

Depending on the time of year you'll find chillies and peppers, tomatoes, artichokes, sunflowers, grains, pulses (peas and beans), fruits including Physalis (Cape gooseberry), loquats, apricots and feijoa and a range of growing systems.

Chillies There are dozens of varieties, from mild to unbearable. Chilli heat is measured in Scoville Heat Units, and the hottest variety found so far is rated at 1,500,000SHU, whereas a basic supermarket green chilli comes in at 1,500. Chillies and sweet peppers are easy to grow – you'll find seeds and plants in the Eden shop.

Artichokes Full of antioxidants, aid digestion, provide roughage and are used by some as a hangover cure. They are also said to lower bad cholesterol and to be good for the liver.

Feijoa (Acca sellowiana) The egg-shaped green fruit tastes like apple, mint and pineapple rolled into one. The flower petals are edible too.

Citrus M.10

The citrus family is fond of breeding. Clementines are a cross between mandarins and bitter Seville oranges, and tangelos the offspring of tangerines and grapefruits. Citrus fruits provide vitamin C and nutraceuticals (more on these in the Core Café in the Core). Citrus oils are used in flavourings, cleaning products, perfume, anti-bacterial agents, CFC substitutes and even fuel.

Tobacco M.11

Tobacco was originally brought to Europe by Columbus. Some heralded it as a miracle medicine centuries before its link to cancer. The tobacco trade was associated with high profit (from tax revenues), smuggling and slavery. Tobacco caused 100 million deaths in the 20th century. If current trends continue, it may cause one billion deaths in the 21st century (WHO). The growing crop depletes soil fertility, requires many sprays and causes deforestation. Tobacco does, however, provide livelihoods for many farmers in the developing world and could potentially have another use – as the 'lab rat' of the plant kingdom. Scientists are developing a vaccine against tooth decay from GM tobacco and are also working on a vaccine for non-Hodgkin's lymphoma, a cancer of the lymph system. Note: This is a summer exhibit only.

Plants for cut flowers M.12

Over 85% of our cut flowers are imported, causing environmental and social challenges but also potential for jobs and a step out of poverty. Check out the label. Seasonal blooms with low 'flower miles' mean you can say it with sustainable flowers. **edenproject.com/shop**

Dionysus – a question of balance

This potent nature god, here depicted as a bull, was associated with horticulture, fertility, wine, festivities, intoxication, illusion and also destruction. Dionysus started out with good intentions as the god of vegetation. However, things changed when he went from growing the vine to drinking its fermented juices ... party time! The land, like Dionysus, has changed. Here he stands, straddled between the ancient landscapes and the irrigated lands of intensive agriculture.

It's possible to go too far but it's possible to regain the balance: people can change things for the better.

Tim Shaw created this wild Bacchanal where dancing Maenads mirroring the shapes of the vines surround their god, Bacchus (aka Dionysus).

Grape vines M.13

Grapes were one of the first cultivated fruits and today are the most widely planted fruit crop worldwide. Our long-term relationship stemmed from our ability to preserve them, by drying and far more consequentially by turning them into an alcoholic drink.

Olives M.14

Once olive oil provided light for lamps and anointed the brave. Now used mostly in the kitchen, it is thought to reduce cholesterol levels and deter heart disease. Production is up but the squeeze is on to reduce chemical inputs. There is as much variation in taste and character in olive oil as there is in wine.

The Med Terrace and vegetable, herb and fruit garden M.15

The classic Mediterranean diet is associated with good health and long life. The diet is based around:

- fresh vegetables (including greens, salads and legumes)
- fruits and herbs
- nuts and cereals
- olive oil (monounsaturated fat)
- a bit of fish, poultry and dairy (yogurt or cheese)
- small amounts of red meat (low intake of saturated fats)
- a little alcohol (usually wine with a meal)

Is it just the food or is it also getting together with friends and family to relax and chat that does you good? Savour the flavour in the Med Terrace café and watch your food growing around you.

Eden Live – Seasonal Events Programme

The seasonal programme for 2015

Eden Live's events programme provides a different Eden every time you visit. There's something for all the family.

Spring: March to May

Wellcome photographic exhibition (19 March–end May)

The Great Eden Egg Hunt Easter holidays (28 March–12 April)

International Male Voice Choir Festival (3–4 May)

Eden Classic sportive cycling event (10 May)

Strange Science May half-term (22–31 May)

Summer: June to August

Green Fingers Festival (1–21 June) with celebrity gardeners* sharing tricks of the trade

(*check website for specific days of talks and demonstrations)

Dinosaurs Summer holidays (24 July–2 Sept)

Autumn: September to November

Harvest (10–27 Sept) with celebrity chefs* sharing tricks of the trade

(*check website for specific days of talks and demonstrations)

Beer festival (3 Oct)

Marathon (18 Oct)

Halloweden October half-term (24 Oct–1 Nov)

Ice rink opens 24 Oct

Winter: December to February

Christmas at Eden (28 Nov–3 Jan) with Christmas in the rainforest

Stick man February half-term (13–21 Feb)

Ticketed events and hire events

- Eden Sessions (**edensessions.com** for bands and full details).
- Private parties: Eden offers a unique venue for private events. From weddings to conferences, parties to team days, large gala dinners to intimate marriage proposals, we can cover it. 01726 811950, **eventsandfunctions@edenproject.com** or Facebook: Eden Project Wedding and Event Venue

There are loads of special treats such as Little Monsters Ball and meeting Father Christmas too. Plus workshops, specialist days such as 'be a gardener for the day', and tours, including our ever-popular chilli and chocolate tour. Contact the **boxoffice@edenproject.com** or ring 01726 811972 for details or to register interest. Details correct at the time of going to print. For full details of all events visit **edenproject.com**

Eden – the back story

Eden, established as one of the Landmark Millennium Projects to mark the year 2000 in the UK, is an international visitor destination featuring spectacular planting and architecture.

We're an educational charity, celebrating our connections with and dependencies on plants and each other. We all know that the 21st century brings many challenges: food security, moving and rising populations, plant and animal extinctions, increasing energy costs, economic shifts – all cranked up by climate change.

What to do? Sit and weep or do something? Eden regenerated this china clay pit as a symbol of transformative change, to demonstrate what people can do when they collaborate and put their minds to something. The world's challenges will demand the best of all of us: our creativity, ingenuity, understanding, science, technology, enterprise, humanity and our ability to work together. It's all possible, humans are pretty resourceful when they are asked to raise their game.

Our exhibits and events celebrate our dependence on plants, and each other, and tell many of the changing stories of plants and people. We work on and share examples of many practical projects, close by and worldwide, that explore new ways of living in the 21st century.

> 'The future is ours to invent. Let's create a world we want to live in.'
>
> Dr Tony Kendle, Creative Director, Eden Project

Your entrance fees and the money you spend during your visit go to support both the operation of the site and our education programmes and projects.

Eden is also a social enterprise, doing business to give the greatest possible benefit to the widest number of people and showing that improving the environment and livelihoods and building stronger communities can work hand in hand. Since 2001 Eden's 13m+ visitors have helped us put over £1.2 billion pounds into the regional economy through year-round trade with local suppliers and businesses.

In a nutshell

The Eden Project connects us with each other and the living world so that we can work towards a better future. The chapters in our story:

- Regeneration/transformation is possible
- We depend on plants and each other
- We need to look after the plants/planet because there are challenges
- It is possible to overcome the challenges
- We share ways of doing things to achieve this

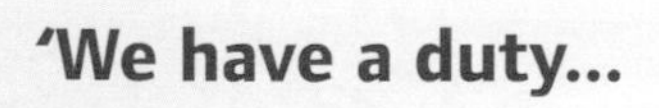

'We have a duty...

...to hope' Barbara Ward

Regeneration and the art of the possible

The beginning of the story:

'Around the late 1990s a small group of people gathered in pubs, hotels and offices to talk about an idea – to create a place like nothing anyone had ever seen before; a place that explored our place in nature, a place that demonstrated what could be done if people who wanted to make a difference got together. It was ridiculous to imagine it was possible and that hundreds of people trained to say no could be persuaded to say yes. But the greybeards had a brilliant plan: ask the youngsters to do it – they don't know it can't be done.' Tim Smit, *Eden*

While Tim Smit (now Sir Tim Smit KBE, Executive Vice Chairman & Co-Founder) was restoring the Lost Gardens of Heligan he realised that plants could be made far more interesting by weaving human stories around them, tales of adventure, emotion and derring-do. There was a big story to be told: plants that changed the world. A summer sunset on a china clay tip conjured thoughts of ancient civilizations in volcanic craters, and of putting the largest greenhouses in the world in a huge hole. Why not? We bought an exhausted, steep-sided clay pit 60 metres deep *(below)*, the area of 35 football pitches, with no soil, 15 metres below the water table, and gave it life: a huge diversity of plants we use every day but often don't get to see, planted in soil made from 'waste' materials, watered by the rain, in giant conservatories and buildings that drew inspiration from nature.

> 'Never underestimate the power of a small group of people to change the world. In fact, it is the only way it ever has.' Margaret Mead

Making it happen...

In 1994 Restormel, our local Borough Council at the time, took a leap of faith and put up the first £25,000, so giving the story a beginning.

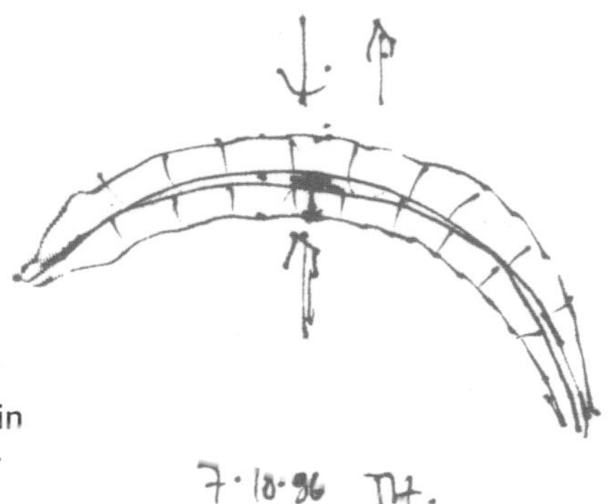

The first Biome sketch, 1996, in the pub, on the proverbial napkin.

A year on, Grimshaws the architects took the baton from Cornish architect Jonathan Ball (co-founder of the Project) and designed our fabulous buildings (at risk, though we paid them in the end!). The McAlpine Joint Venture worked for 18 months without payment or contract and then loaned Eden a significant sum only to be repaid if the Project was successful. This risk-sharing broke down the traditional barriers between designers and contractors and created a team dedicated to one vision. Why did they do this? Because they wanted to change something and because they wanted to say, 'I'm glad I did,' rather than, 'I wish I had.'

Sound simple? Not really. We were turned down by the Millennium Commission (MC) the first time we applied, and many left good jobs before we had raised a bean – or found a site. When our reworked bid secured £37.5m from the MC (huge thanks, MC), we had to match it. For the next 5 years a small team worked tirelessly (mainly in a shed) to turn the idea into a plan and then into reality. Money was raised, fledgling teams grew thousands of plants, mapped them on to the site, started planning the stories ... We recruited a team to run the place and made sure that as well as having a good idea and a fabulous theatre we had the ability to operate it. The Visitor Centre opened in 2000 so the public could watch the construction and share the adventure. The whole site opened on 17 March 2001. For more, read Tim Smit's *Eden* and visit **edenproject.com**.

The recipe for Eden

- Take an exhausted, steep-sided clay pit 60m deep, the area of 35 football pitches, with no soil, 15 metres below the water table.
- Carve the pit into a flat-bottomed bowl and landscape the sides.
- Mix and add 83,000 tonnes of soil made from recycled waste.
- Add superb architecture that draws inspiration from nature to remind us of human potential.
- Colonise with a huge diversity of plants, many that we use every day (but don't often get to see).
- Harvest the water draining into the pit and use it to irrigate our plants (and flush the loos!).
- Season with people from all walks of life.
- Spring 2000, open for a preview of the making.
- Spring 2001, open and serve.

Building our Living Theatre

Massive sandpit to global garden

To make the pit suited to people rather than mountain goats 17 metres was sliced off the top and put in the bottom. 1.8 million tonnes were shifted in 6 months. Dodgy slopes were shaved to a safe angle and terraces created. 2,000 rock anchors (some 11 metres long), stabilised the pit sides. A plant seed soup was sprayed on the slopes to knit the surface.

The answer lies in the soil 83,000 tonnes of soil was made. Minerals came from mine waste (sand from Imerys china clay and clay from WBB Devon Clays Ltd). In the Biomes, composted bark provided long-lived organic matter. The Rainforest Biome plants needed a rich organic soil that could hold water and nutrients, while the slower growers in the drier Mediterranean Biome used a sandier mix. A nutrient-free mix was used in South African Fynbos, where fertile soil is toxic to some native plants. Outdoors, we used composted domestic green wastes. Worms were added to help dig and fertilise.

The Biomes: building the world's largest conservatories

Building 'lean-to greenhouses' on uneven surfaces is tricky. 'Bubbles' were used because they can settle on any shaped surface.

Overall design Two-layer, curved space frame, 'the hex-tri-hex', with an outer layer of hexagons (the largest 11 metres across), plus the odd pentagon, and an inner layer of hexagons and triangles bolted together. The steelwork weighs only slightly more than the air contained by the Biomes. They are more likely to blow away than down, so are tied into the foundations with ground anchors (giant tent pegs).

Transparent foil 'windows' Ethylenetetrafluoroethylenecopolymer (ETFE): three layers, inflated 2-metre-deep pillows, lifespan over 25 years, transmit UV light, non-stick, self-cleaning. They weigh less than 1% of the equivalent area of glass, but can take the weight of a car. We got into the *Guinness Book of Records* for using the most scaffolding, 230 miles of it – sorry to anyone who was needing some that year.

The ETFE pillows were installed by 22 professional abseilers – the sky monkeys.

Bring on the plants – extreme gardening

Our Green Team often plant on near-vertical banks; have planted millions of plants of around 5,000 types; plant over half a million bulbs every autumn; plant around 60,000 new plants annually; start at 7.30 every day to prune and water before you arrive; do 50 hours weeding a week in the summer; remove about 25m^3 of green waste from the site every week; recycle this material to make over 120 tonnes of compost a year.

Plants Many are grown from seed in our nursery, some come from botanic gardens, research stations and supporters, mostly in Europe and the UK. We keep a record of every plant on site. Pollination: some plants are insect-pollinated, some wind-pollinated, some paintbrush-pollinated! We do this when we need the flowers to produce seeds.

Pruning Rainforest trees are pruned by abseilers, tree climbers and from a cherry picker.

Pest and disease control Our rigorous healthcare programme starts with isolation houses at Eden's nursery to catch problems before they reach the site. On site our integrated pest management system uses cultural methods (removal of infested plant parts), 'soft' chemicals (soaps and oils) and biological control (bugs that eat bugs). Spot the little bamboo pots on pulley systems that give the bugs a ride to the treetops. Also eating their fill of pests are lizards, birds (e.g. Sulawesi White Eye) and frogs (e.g. White's Tree Frog). Our UV lightboxes catch pests and monitor their numbers.

Climate In the covered Biomes it is monitored and controlled automatically. The main heating source is the sun; the back wall acts as a heat bank, releasing heat at night. Triple-'glazed' windows provide insulation, while air-handling units cool on hot days and heat on cool ones.

Rainforest Biome Kept between 18-35°C and up to 90% humidity.

Mediterranean Biome 9°C min. in winter, 25°C max. in summer. Drier than the rainforest; even in cooler periods the vents often open to reduce humidity and fungal problems.

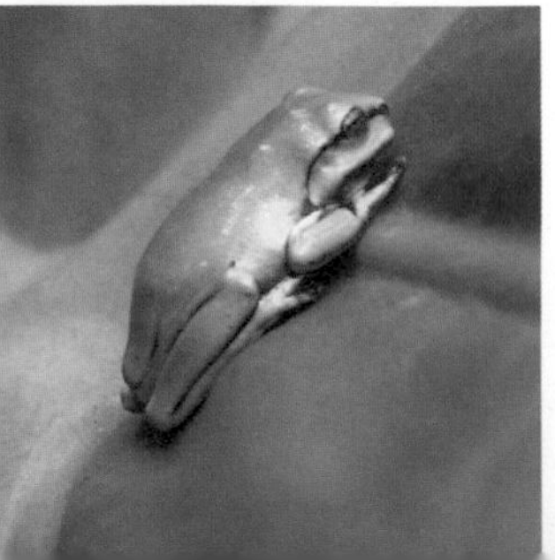

Bring in the people

Since opening in 2000 we've developed gardens, exhibits and exhibitions, events, workshops, courses, education programmes (for all), concerts, sustainable buildings, efficient operating, energy and waste systems and grown our team from 5 staff to 460, plus 134 volunteers. Teams report to the Board, who report to the Trust, who ensure the operation meets its charitable aims. Art, science, horticulture, education, management, retail, catering, philosophy, economics, design, construction, publishing, research, housekeeping, guiding, fundraising, storytelling, marketing, media – between them the teams cover all these bases and more. We work with others whenever we can to discover new approaches and to share what we have learnt. Full details of Eden's operational structure can be found at **edenproject.com**.

We explore creative approaches to physical access and information sharing to ensure our work is accessible to all. See **sensorytrust.org.uk** for more information.

Eden is a project, a living laboratory exploring new ways of doing things. The way we operate and the projects we run all test new ways of living. We take people, the planet and profits with a purpose into account ('the triple bottom line'). This begins with building an awareness of what we dare not lose and exploring what we must work with to help create a viable society: social and environmental change, carbon reduction, energy security, food security, conservation, building communities that can cope...

We love stories

Story is a powerful means of captivating, providing insight, testing moral choices, painting possible futures, challenging and holding a mirror up in a way that is acceptable – personal and impersonal at the same time. Unless a culture has strong stories it loses its direction. Eden aspires to be a place where the stories of our future are created and told – the *Aesop's Fables* of the 21st century. Look out for Eden's storytellers, who will share their stories with you.

Eden Learning

eden project LEARNING

Schools

Our schools programme, for all key stages, hosts over 45,000 young people a year. We aim to help provide context, purpose and meaning. Programmes are designed to help teachers meet curriculum targets across a range of subjects, from geography, art and science (biodiversity, biology, botany, conservation, ecology, and horticulture) to architecture and sustainability. You may spot young explorers in 'Rainforest Uncovered', reconnecting with science on the ultimate field trip or students seeking mathematical patterns in 'Finding Fibonacci'. Working with the YHA we also offer accommodation and school trip adventures: Rainforest Recruits and Planet Eden. (**yha.org.uk/school-trips/hostel/eden-project**). Each year Eden trains hundreds of teachers to make best use of their outdoor classrooms for learning and play and we also visit schools to run workshops on sustainability and enterprise. We love to use Eden as a venue to celebrate young people's creativity and talent – local schools provide us with the best-dressed Christmas trees in the county – and we support national events like Empty Classroom Day in June each year.

Further and Higher Education

Eden welcomes over 10,000 further and 1,000 higher education students annually across a range of subjects including continuing professional development for teachers and a Special Study Unit for local medicine students called 'From Plant to Pill'. Our MSc Sustainability (Working for Positive Change) in collaboration with Anglia Ruskin University and Change Agents UK takes students on a unique learning journey at Eden and Cambridge, preparing them to become successful change agents within the world of work. **www.mscsustainability.org**

New for 2015

From September 2015 the Cornwall College Group in partnership with Plymouth University will be running Higher Education courses with Eden at Higher National Certificate, Higher National Diploma, Foundation Degree and BSc degree level including:

- FdA Event Management
- HND Performance, Storytelling and Interpretation (subject to validation)
- FdSc Horticulture
- HNC, HND and BSc (Hons) Horticulture in Garden and Landscape Design
- BSc (Hons) Horticulture (Plant Science)
  cornwall.ac.uk/eden-degrees

Apprenticeships

Eden have partnered with Cornwall College Group to deliver bespoke two-year apprenticeship programmes in a number of career streams. Key areas comprise Eden Chef, Eden Gardener and Eden Host. We also host Eden Apprentices in other areas such as HR, Media and PR, Graphics, Vehicle Maintenance and Finance. Our in-depth programmes offer an education in sustainable business as well as the chosen career path.

Eden Leadership Experiences

Eden is also engaged with the business world as it represents a huge potential force for change. We run innovative leadership experiences for individuals and businesses, both large and small, who are committed to shaping a better future.

In addition to all the above we run a whole range of workshops, leisure learning, community camps and team-building away days. We also offer accommodation on site through the YHA (see p.59).

Visit **edenproject.com/learn-with-us** for more information on all these programmes.

Eden's projects

Eden works on a range of projects that explore how people can work together and with nature towards a better future.

The Big Lunch

The Big Lunch is the UK's annual get-together for neighbours, providing the perfect recipe to have fun, feed community spirit and help to build stronger neighbourhoods. People are the key ingredient, with those taking part creating friendlier communities in which they start to share more, from conversation and ideas to skills and resources. At a Big Lunch small talk often leads to big talk, and together people start to tackle local issues; from reducing crime or feelings of isolation. 84% of people who took part in 2014 said it helped them feel better about their neighbourhood and 97% would recommend taking part to their family and friends. The Big Lunch is on the first Sunday in June every year. Have one! Request a pack at **thebiglunch.com**

Big Lunch Extras

We believe communities often hold the answers to what a better future looks like. Sometimes all they need is a little help along the way to unlock ideas, make things happen and get people involved. Big Lunch Extras builds on the success of the Big Lunch and supports individuals with the skills, confidence and motivation they need to create positive change where they live. **www.biglunchextras.com**

The Big Lunch and Big Lunch Extras are made possible by the Big Lottery Fund.

Gardens for Life

In Africa, being able to grow your own food can make a real difference to your prospects. The Eden Project's Gardens for Life programme works with schools and communities in Kenya and the Gambia to feed minds and bodies, giving them essential skills for the future and helping ensure that everyone gets at least one square meal a day – using ingredients they've grown themselves. Free school meals helps ensure higher attendance rates and improves grades for students, but a whole community that can grow its own food ensures that everyone benefits. **edenproject.com/gardens-for-life**

Gardens for Life

Working with others

We also provide a platform to tell stories of projects from like-minded people we've met along the way such as Fair Wild; sustainable harvesting of wild herbs (p.14), wild rubber-tapping to support livelihoods and conserve rainforests (p.28) and supporting livelihoods in Africa through baobab harvesting (p.30).

We work closely with other organisations:

People and Gardens was set up to enable people with learning disabilities or emotional impairments to be able to develop as individuals and to have equality of choice and opportunity in the workplace. The founders of People and Gardens understand through their own experiences that we should all work together to break down barriers, to educate and to support each other to make the world a better place for everyone. **peopleandgardens.co.uk**

The Sensory Trust makes places more accessible, attractive and useful for everyone, regardless of age, disability and social circumstances. Their work brings social and health benefits to people whose lives are affected by social exclusion, including older people, people with physical, sensory and intellectual impairments and families and carers. **sensorytrust.org.uk**

Our journey continues, working with others, exploring new ways of doing things in education, in business and in communities.

Ways of working in the 21st century

We're committed to reducing our carbon footprint, using resources efficiently, generating our own renewable resources and being self-sufficient in soil, water and energy where possible. Our tens of thousands of plants sequester carbon every day. We also operate a green travel plan. You can get involved too: see **edenproject.com**.

Energy efficiency The Biomes' hexagons copy nature's honeycombs: maximum strength, minimum materials. All our buildings set high standards for good building design and process, and demonstrate the worth of natural materials and structures. Since 2008 we've reduced our emissions from gas use by 20%, and electricity by 34%. New control systems, boilers, lighting, etc (thanks partly to an interest-free loan from the Carbon Trust), should see a further 8% reduction in CO_2 emissions relating to electricity and gas since last year.

Waste Neutral We reduce, re-use and recycle our waste wherever possible (currently recycling 20 different waste streams) and we reinvest by purchasing items that are made from recycled materials. Since 2005 our in-vessel composter has turned 196 tonnes of food waste, 29 tonnes of green waste and 23 tonnes of cardboard waste into 127 tonnes of plant compost for use on-site.

Cleaner technologies Photovoltaic panels installed on the Core roof, our workshops and warehouse buildings, together with a small wind turbine, generate clean electricity. Nearly half our water needs – averaging 20,000 bathfuls a day – are provided from harvested ground and rainwater on site by a subterranean drainage system. We use this resource to irrigate our plants and flush our loos.

> 'You never change things by fighting the existing reality. To change something, build a new model that makes the existing model obsolete.' R. Buckminster Fuller (1895–1983), architect, designer, visionary

Project: Geothermal

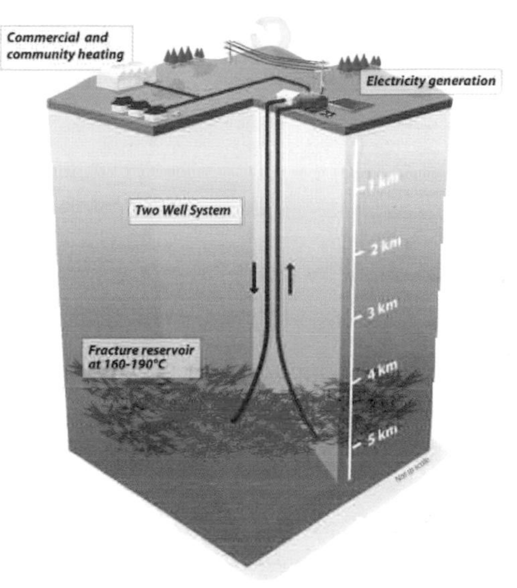

With EGS Energy, we've got planning permission for a deep geothermal plant to supply us with renewable heat and power, and export the excess electricity, enough for around 3,500 households. This new technology, if rolled out, has the potential to supply as much as 20% of the UK's electricity and all of its heating! More on this at **edenproject.com/geothermal**, and on our overall energy policy and action plan at **edenproject.com/sustainability**. EGS and Eden are in discussions with the government and commercial lenders to find the required funding.

Food and retail

From plant to plate: our sourcing policy means that the food we serve in our cafés reflects the stories we tell in our exhibits. It is responsibly sourced: fairly traded, single-source, organic, seasonal and/or local and freshly made, often in front your eyes. We cater for vegetarians, vegans, meat-eaters, gluten-free diets. Check site for opening times of venues (some close in quieter periods).

We work with local growers, suppliers and companies to develop products for sale. Eden's catering uses local produce wherever possible; 91% is bought in Cornwall and Devon. Low food miles, seasonal and vegetarian dishes all help reduce greenhouse gas emissions. Fairtrade, organic and other certified products from further afield demonstrate that good trade is a vital part of sustainability. The products you find in our shop all have a story to tell and are sourced with the planet in mind.

Where has the money come from?

The Millennium Commission weighed in with £37.5m of lottery funding to single Eden out as the 'landmark' project of the far South West, and their subsequent contributions brought the total to just over £56m.

We hope we've delivered for them and for anyone who ever bought a lottery ticket. Other major sources of funding included the EU and Southwest Regional Development Agency (some £50m between them) and £20m of commercial loans. The balance was made up of other loans and some funds generated by Eden itself and reinvested back into the Project.

Maintaining a strong and diverse financial base is crucial to preserving the Eden Trust's independence and credibility. A full list of all our funders to date can be found at **edenproject.com/thankyous**. Thank you.

How much?	£m
Buying a large and unusual site, car parks, roads and paths	16
Reshaping the ground to make it safe, dry and useful	8
A couple of decent greenhouses	26
40 acres of plants ... some tall	3
83,000 tonnes of manufactured soil to grow them in	2
A nursery to practise in and grow some unusual plants	1
Buildings for you and our team – fully equipped	22
Services to keep it all running	7
Paying the team up to opening	3
Exhibits to entertain you, walkways to keep you dry, a lift, a bridge	12
Advice on the things we couldn't do ourselves	12
Investments in our future like the Foundation building	9
A spectacular home for education – The Core	16
Warehouse, gatehouse, waste compound, Arena	4
Dreams cost money — Total	141

Get involved

Eden is a charity and our work and successes are only possible thanks to the generosity of our members, donors and volunteers.

You can support our work in many ways

Visit us You're helping just by being here – all the profits from your visit go to the Eden Trust.

Gift Aid your admission fee This allows us to claim 25p back from the taxman on every pound you give.

Join us Becoming an Eden Member is a fantastic way to support our charitable work and learn more about us. Whether you're into gardening and good food or want to support our environmental and educational projects, joining us means you become part of Eden. To thank you for your support you will receive free entry to Eden for you and an adult family guest on every visit, our quarterly Eden magazine, invitations to exclusive events and monthly e-newsletters packed with interesting updates and horticultural tips.

Please visit **edenproject.com/membership** or call 01726 811932 to find out more.

Donate The money we raise goes towards our work on public and formal education, research, conservation and sustainable futures. **edenproject.com/donate**

Volunteer Can you spare at least 6 hours a fortnight? Feel good, learn things, meet people – what's not to love? An ideal way to become involved in the daily work of the Project. **edenproject.com/careers/volunteering**

Subscribe to our free newsletter **edenproject.com/e-newsletter**

Learn with us Discover more about our learning opportunities on pp.52 3

Useful numbers

General enquiries: 01726 811911 Box office: 01726 811972
Group bookings: 01726 811903 School bookings: 01726 811913

The Eden Project, Bodelva, St Austell, Cornwall PL24 2SG
The Eden Project is a registered charity no. 1093070

Contact us **edenproject.com/contact-us**

What can we do?

There's no shortage of top tips for a greener life: fitting energy-saving lightbulbs and so on. Useful? Yes and no.

They can trivialise the issues; saving the world isn't a matter of what goes in your basket, and it takes the heat off the big guys. But if we had to write our own Ten Top Tips (eleven, actually!), they'd go like this:

1. Do stuff Don't waste it, turn it off, turn it down, do it less, do it local, do it yourself, recycle, swap, repair, share.

2. Be hopeful Hope isn't just about crossing your fingers. Without it we could get cynical and frozen in despair. Hope is the fuel – but it only works if you do something.

3. Learn about your life Is having 'stuff' bad? Not always: trade is not the same as consumption and can support livelihoods. Understand what sustains you and what you need to care about. Learning new talents and skills can help you get there.

4. Increase your reach There's only so much you can do on your own. Try working with or through other organisations. Also don't forget that your wallet is your weapon. Make buying choices that help good things happen – worldwide.

5. Be angry at the things you can't change ... but think about who can change them. Demand that governments, companies and big organisations change with us and give us real choices.

6. Imagine different things The 21st century will be a time of transformation. Meet different people, explore different things, read different books, try out new ideas.

7. Give gifts and give thanks
Understand why we need each other. This is a time to support each other, to work together and build communities.

8. Get out more People can't care about what they don't understand and don't have some sense of connection to. So we need to get out and down in that dirt lest we forget how it keeps us alive. Play together, learn, explore and have adventures.

9. Forgive yourself (and others)
Sustainable development will be a territory for endless exploration. Learn from mistakes. We make mistakes because we act, strive and aim high – and that is what makes us human.

10. Have fun 'Living a sustainable life' isn't all about 'don't do this' sucking the joy out of living. Where is the adventure in that? There are worlds of possibility out there. Rich cultures, rich experiences, music, laughter, fun and just enjoying life more – foundations for a better future!

11. Be the change you wish to see in the world Gandhi's saying sounds like something from a hippie poster, but actually it was one of the greatest social insights of the 20th century. So, do everything positive you can, not because a list has told you to but because it's who you want to be.

First published 2001 by Eden Project Books, a division of Transworld Publishers

Fifteenth revised edition 2015

Text by Dr Jo Elworthy with assistance from the Eden team

Transworld Publishers, 61–63 Uxbridge Road, London W5 5SA,
a division of the Random House Group Ltd

booksattransworld.co.uk/eden

ISBN 978 1909513044

Editor: Mike Petty Design: Charlie Webster Printed in Great Britain

Cover photo by Claire Braithwaite

Many fine photographers, whether professionals, Eden colleagues or Eden friends, have contributed to this edition. Sadly, limited space doesn't enable us to list them all individually, but we thank them for their help.

Penguin Random House is committed to a sustainable future for our business, our readers and our planet. This book is made from Forest Stewardship Council® certified paper.

The Eden Project is owned by the Eden Trust, registered charity no. 1093070 and all monies raised go to further the charitable objectives.

Eden Project, Bodelva, St Austell, Cornwall PL24 2SG

T: +44 (0)1726 811911 F: +44 (0)1726 811912

edenproject.com

Supported by **The National Lottery®** through the Millennium Commission

Welcome to the Rainforest Biome

R.01 Tropical islands
R.02 Southeast Asia
R.03 West Africa
R.04 Rainforest Canopy Walkway. Phase 1
R.05 Tropical South America
R.06 Welcome to crops and cultivation
R.07 Growing with the forest
R.08 Regrowing the forest
R.09 Rubber
R.10 Cocoa and chocolate
R.11 Stories from the rainforest
R.12 Palms
R.13 Tropical fruits – Bananas and friends
R.14 Sugar
R.15 Rice
R.16 Tropical fruits – Baobab and friends
R.17 Bamboo
R.18 Coffee
R.19 Nuts and spices
R.20 Secrets of the rainforest

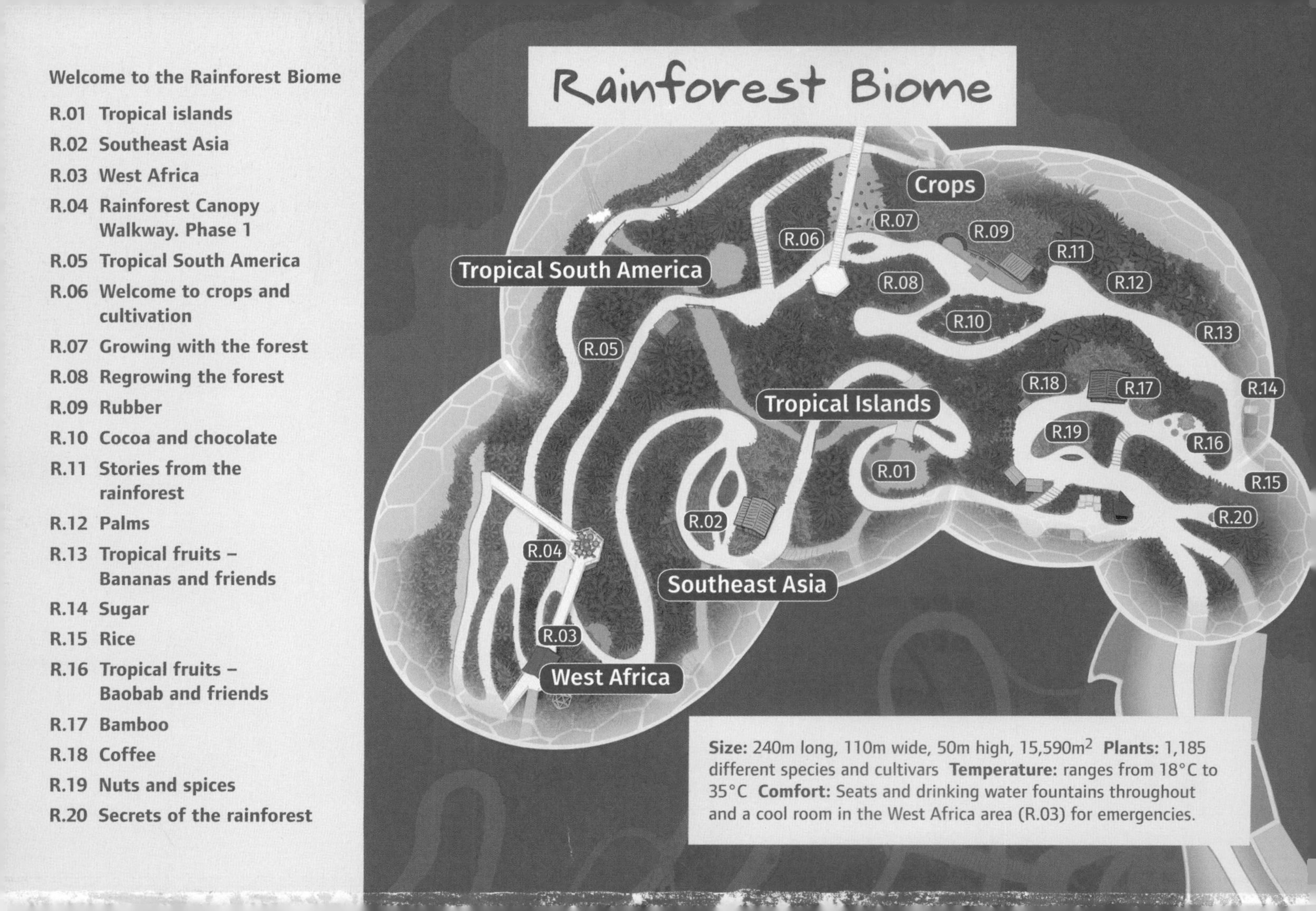

Size: 240m long, 110m wide, 50m high, 15,590m^2 **Plants:** 1,185 different species and cultivars **Temperature:** ranges from 18°C to 35°C **Comfort:** Seats and drinking water fountains throughout and a cool room in the West Africa area (R.03) for emergencies.